YOUR KNOWLEDGE HAS VALUE

- We will publish your bachelor's and
  master's thesis, essays and papers

- Your own eBook and book -
  sold worldwide in all relevant shops

- Earn money with each sale

Upload your text at www.GRIN.com
and publish for free

Animal Genetics Erosion, Endangerment, Extinctions and Conservation Strategies

Feyisa Lemessa

Bibliographic information published by the German National Library:

The German National Library lists this publication in the National Bibliography; detailed bibliographic data are available on the Internet at http://dnb.dnb.de.

ISBN: 9783346943767
This book is also available as an ebook.

© GRIN Publishing GmbH
Trappentreustraße 1
80339 München

All rights reserved

Print and binding: Books on Demand GmbH, Norderstedt, Germany
Printed on acid-free paper from responsible sources.

The present work has been carefully prepared. Nevertheless, authors and publishers do not incur liability for the correctness of information, notes, links and advice as well as any printing errors.

GRIN web shop: https://www.grin.com/document/1394630

HARAMAYA UNIVERSITY

SCHOOL OF GRADUATE STUDIES

REVIEW ON ANIMAL GENETICS EROSION, ENDANGERMENT, EXTINCTIONS AND CONSERVATION STRATEGIES

FEYISA LEMESSA

College: College Of Agriculture and Environmental Sciences

Department: Animal and Range Science

Program: Animal Breeding and Genetics

March 21, 2022

Haramaya University, Haramaya

Contents

Introduction

The world watch list for domestic animal diversity documents more than 6 300 breeds (FAO, 2000) and 7616 (L. F. Groeneveld, J. A. Lenstra, H. Eding, M. A. Toro, B. Scherf, D. Pilling, et al. 2010; A. S. Mariante *et al.,* 2009) breeds of livestock belonging to 30 domesticated species. These breeds were developed following domestication and natural and human selection over the past 12 000 years. The current number of breeds is likely an underestimation since a large proportion of indigenous livestock populations of the developing world, where most animal genetic resources are found today; have yet to be described at phenotypic and genetic levels. The predominant species include cattle, sheep, goats, pigs, chickens, horses and buffalo [**Hoffmann, 2010**]. Several other domesticated animals like camels, donkeys, elephants, reindeer and rabbits are also valuable to different regions of the world [S. Anderson, 2003]. Cattle, sheep, chickens, are predominantly found all over the world, while goats and pigs are less uniformly distributed. Livestock populations have evolved a unique adaptation to their agricultural production system and agro ecological environments. Their genetic diversity has provided the material for the very successful breeding improvement programmes of the developed world in the 19th and 20th century. This represents a unique resource to respond to the present and future needs of livestock production, both in developed and developing countries. Recognizes The well importance of animal production for food security and rural livelihoods and particularly AnGR (defined by FAO, 2000 - Box 1.1) supply over 30% of total human requirements for food and agriculture (FAO, 1999) and contribute to the livelihood of 70% of the world's rural poor (LID, 1999). However, the purposes of raising livestock go beyond their direct output functions and include other significant economic and cultural roles. These include savings, insurance, cyclical buffering, accumulation and diversification, as well as various cultural roles related to status and the obligations of their owners (Anderson, 2003).

However, livestock diversity is shrinking rapidly (of which about 30% of them are at risk of extinction (A. S. Mariante, M. S. M. Albuquerquea, A. A. Egito, C. McManus, M. A. Lopes, and S. R. Paiva, 2009). With the exception of the wild boar, which is the ancestor of the domestic pig, and wild red jungle fowl, which is the ancestor of the domestic chicken, the putative wild ancestors of our major livestock species, the reservoir of genetic diversity, are now either extinct (e.g. the auroch, the wild ancestor of cattle, or the ancestral species of the Old World camelids) or low in numbers and threatened by extinction (e.g. wild goat populations of the Near East, vicuña from Andean plateau, wild donkey in Africa). Among the domesticated populations, it is estimated that one to two breeds are lost every week (**FAO, 2000**). However, the impact of these losses on global or local diversity remains undocumented. While it is already too late for many breeds in Europe, the situation is also particularly worrying in the developing world where rapid changes in production systems are leading to the replacement of breeds or at best crossbreeding. There is therefore an urgent need to document the diversity of our livestock genetic resources and to design strategies for their sustainable conservation and the main objective of this review is to exploring

2. LITERATURE REVIEW

2.1 Livestock and Biodiversity

Biodiversity is considered to form the very basis of life on earth. The importance of biological diversity is for evolution and for maintaining life sustaining systems of the biosphere." They affirm that "the conservation of biological diversity is a common concern of humankind" and that they are "aware that conservation and sustainable use of biological diversity is of critical importance for meeting the food, health and other needs of the growing world population." The Conference of the Parties (COP) Decision IV/10 acknowledges that "economic valuation of biodiversity and biological resources is an important tool for well-targeted and calibrated economic incentive measures.

Consistent with this is a view of biodiversity as insurance in the face of uncertainty. Because the value of biodiversity is not completely captured in markets, its conservation is often at loss in comparison to land-use developments for market goods and access. As some papers have recently outlined (e.g. Costanza et al., 1998; Norton et al., 1998), the natural capital is the very foundation of economics. In this perspective, valuing biodiversity is important because it makes markets and economics commensurate with the actual functioning of the world.

On the other hand, with regard to specific to livestock biodiversity, Livestock species are unlikely at danger of extinction themselves. The level of biodiversity which is of concern here is that of breeds and even populations within breeds. In fact, within- breed diversity account for 50 to 70% of total genetic variance (Hammond and Leitch, 1996). In the case of livestock, the anthropogenic impact on biodiversity is often the most important one, through controlled reproduction and selection, as well as introgression and production decisions impacting the demography of livestock populations. Measures of diversity used in the context of wild species are hence often not suitable as measures of diversity for livestock breeds. Therefore, Livestock biodiversity is less implied in the functioning of complex food chains and ecosystems than wildlife biodiversity. However, AnGR generally have implications for cultural and landscape diversity and are of great significance for the resilience of many agro ecosystems. Finally, domesticated animals have primarily direct use values as opposed to wildlife, though this distinction is not as strictly dual as it may appear. In consequence, previous studies have identified that methods used for the evaluation of farm AnGR need to consider their importance for production systems, agro-ecosystems and diversity as specific for livestock breeds.

2.2 The Importance and values of Animal Genetic Resources [AnGR]

More than 90% of exports originate from developed countries, and the share of trade in genetic material from developed to developing countries increased from 20% in 1995 to 30% in 2005 (Gollin *et al.* 2008). In many cases, the improved components of the high-input management systems needed to express the genetic potential of the high-output breeds have been transferred to developing countries. Industrial systems utilizing sophisticated technology and based on internationally sourced feed and animal genetics already produce 55% of pork, 68% of eggs and 74% of poultry meat globally (FAO, 2003; Steinfeld *et al.* 2006).

Economic Values of Animal Genetic Resources in two circumstances have served to increase the importance of establishing the value of genetic resources (Zohrabian et al., 2003). The first is that in the face of strengthening laws for intellectual property protection of germplasm, the past collegial system of free exchange among researchers is breaking down, requiring that some compensation mechanism for genetic sources emerge. The second event refers to the signature in 1992 of the International Convention on Biological Diversity (CBD). Indeed, the first article of the Convention stipulates what follows: The objectives of this Convention, to be pursued in accordance with its relevant provisions, are the conservation of biological diversity, the sustainable use of its components and the fair and equitable sharing of the benefits arising out of the utilization of genetic resources, including by appropriate access to genetic resources and by appropriate transfer of relevant technologies, taking into account all rights over those resources and to technologies, and by appropriate funding (Article1,CDB).

Mendelsohn (1999) evokes at least three arguments which plead in favour of the economic valuation of AnGR (schematic figure 1): the principal question is to identify the most important attributes. Stated differently, should the breeders focus on fecundity, weight gain, disease resistance and/or other attributes, Human preferences regarding genetic traits of livestock differ across regions, countries, communities and production systems. For example, in marginal areas, the most valuable livestock attributes are often those that successfully guarantee multifunctionality and flexibility or resilience in order to deal with variable environmental conditions. In contrast, in zones with high production potential, livestock attributes maximising productivity of specific products under controlled conditions are more valuable. Thus, there is no single answer to such question. However, if breeders could better estimate the value assigned to developing livestock species for particular production systems, they could make better decisions on which breeds to utilise for intervention. This would also help geneticists design better programmes of sustainable intensification, and it would help decision-makers in determining the effort allocated to such programmes. The Summary of Economic valuation of AnGR has illustrated below

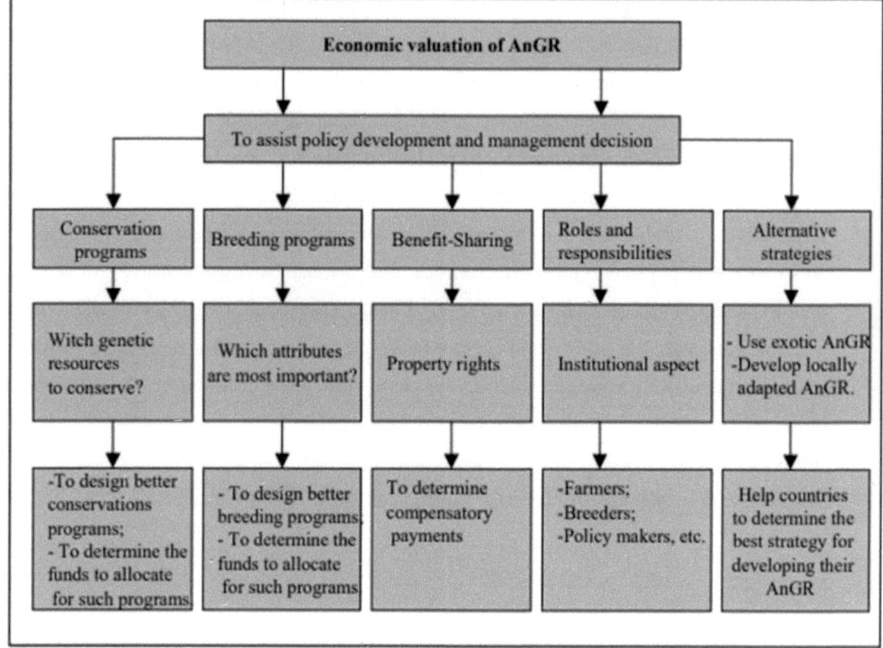

Figure1. Schematic summary of arguments to economic valuation of AnGR

2.3 Status and Trends of AnGR erosion and endangered Species

FAO's latest global assessment of breed diversity identifies 7040 local breeds (each reported by only one country) and 1051 transboundary breeds (each reported by several countries) (FAO, 2009a). A breed is a cultural rather than a biological or technical entity (Eding 2008). A breed covers groups of animals having similar characteristics that depend on geographical area and origin. Most European breeds are well defined, distinct and for a large part genetically isolated. By contrast, Asian and African breeds most often correspond to local populations that do not differ markedly. About two-thirds of reported breeds are currently found in developing countries. Local breeds are commonly used in grassland-based pastoral and small-scale mixed crop-livestock systems, where they deliver a wide range of products and services to the local community, with low to medium use of external inputs. They are usually not well characterized or described, and are seldom subject to structured breeding programmes to improve performance. So-called 'international transboundary breeds' of the five major livestock species (cattle, sheep, goats, pigs and chickens), many of them high-output commercial breeds, have spread globally for use in large-scale, often landless, production systems, where they produce single products for the market (either milk, meat or eggs) with high levels of external inputs.

However, the risk status of 36% of breeds is unknown (FAO, 2009a). The loss of within-breed diversity is not known, although within commercial breeds, high selection pressure, particularly when combined with poor breeding programmes, leads to a narrowing genetic base. About 9% of reported breeds are extinct and 20% are currently classified as being at risk. The species involved in production and marketing systems with fast structural change show high proportions of breeds currently at-risk and already extinct. This includes 38% of chicken breeds, 35% of pig, 33% of horse, and 31% of cattle breeds (FAO, 2009a). Economic and market drivers also made up 28.5% of all responses in FAO's questionnaire survey on threats to animal genetic resources across the main species (FAO, 2009c). It can be expected that multifunctional local breeds continue to play a role in the livelihoods of poor people and in marginal areas.

2.4 Status of endangerment of breeds and indicators

At present, the most widely reported indicators pertinent to livestock biodiversity are found in the list provided by FAO through the "Domestic Animals Diversity – Information System (DAD-IS) and the Animal Genetic Data Bank of the European Association for Animal Production (EAAP). The EAAP data bank monitors information on populations at breed and country levels to keep an eye on development, on the change in the risk of breed extinction and to encourage use and conservation of animal genetic diversity. The farm animal species concerned are buffalo, cattle, goat, sheep, horse, ass, pig, and rabbits. In total, the data bank concerns 46 EAAP member countries and other European countries. Over the past decade, the FAO has helped collecting data from some 170 countries on almost 6,500 breeds of domesticated mammals and birds: cattle, goats, sheep, buffalo, yaks, pigs, horses, rabbits, chickens, turkeys, ducks, geese, pigeons, even ostriches. The FAO Global Databank for Farm Animal Genetic Resources (DAD-IS) contains information on 6,379 breeds of 30 mammalian and bird species. Population size data is available for 4,183 breeds of which 740 breeds are already extinct and 1,335, or 32%, are classified at high risk of loss and are threatened by extinction. DAD-IS monitors breeds worldwide and classifies them into seven risk categories: extinct, critical, endangered, critical-maintained, endangered-maintained, not at risk, and unknown. "Extinct" indicates that it is no longer possible to recreate the breed population. Extinction is absolute when there are no breeding males (semen), breeding females (oocytes), nor embryos remaining. "Critical" indicates that the total number of breeding females is less than 100, or the total number of breeding males is less than or equal to five, or the overall population size is close to, but slightly above, 100 and decreasing, and the percentage of pure-bred females is below 80 %.

"Endangered" indicates that: the total number of breeding females is between 100 and 1000; or the total number of breeding males is less than or equal to 20 and greater than five; or the overall population size is close to, but slightly above, 100 and increasing and the percentage of pure-bred females is above 80%; or the overall population size is close to, but slightly above, 1000 and decreasing, and the percentage of pure bred females is below 80%. "Critical-maintained" and "endangered-maintained" refers to breeds being maintained by an active public conservation programme or within a commercial or research facility. "Not at risk" indicates breeds for which the total number of breeding females and males is greater than 1000 and 20 respectively; or the population size approaches 1000 and the percentage of pure-bred females is close to 100 %, and the overall population size is increasing. Finally, "unknown" covers breeds for which no data are available. Figural status and the collection of data on the above parameters and their distribution across the Global are indicated below (Sourced: FAO (2003) and the report by Roosen, Jutta; Fadlaoui, Aziz; Bertaglia, Marco (2003); Global status' report)

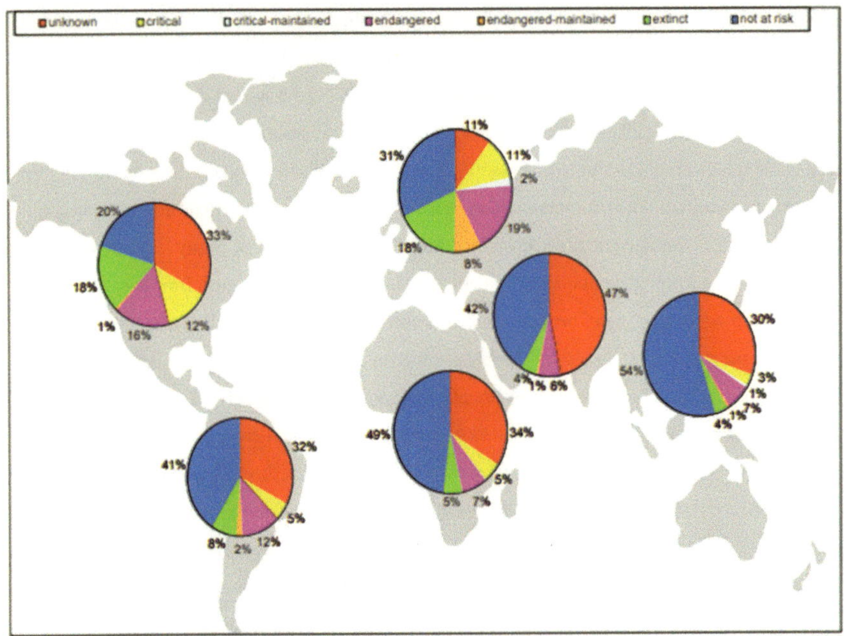

Figure2. Global summary – Proportion of breeds recorded for all species in each region by risk status category as of (FAO, 2003). The threat to farm animal biodiversity is displayed in above figure I. which shows a summary of the status of the world's farm animal breeds. In Europe, the situation of farm animal biodiversity is particularly critical: 18% of the breeds existing in the early 20 th century have already been lost. Unless significant changes take place in the driving forces behind biodiversity depletion, 40% of recorded breeds risk becoming extinct over the next 20 years. This figure illustrates the reality that erosion of biodiversity at the breed level is not simply a concern for the distant future, but an active ongoing process

2. 5 the driving factors for AnGR erosion and the consequences

The global livestock sectors consist of a variety of production systems. In extensive systems in developing countries, animals are often used for a variety of functions, while in intensive systems the focus is primarily on food production. Livestock has been undergoing constant genetic change, which is the normal state for AnGR. Breed development is a dynamic process of genetic change driven by environmental conditions and selection by humans. Natural and artificial selection, crossing between stocks and replacement of one stock with another stock are inherent features of livestock production systems. Global exchange of AnGR has been vital for breed and livestock sector development at all continents. Gibson & Pullin (2005) describe several phases of livestock breeding in industrialised countries. In the 19th century, urbanisation and the development of more intensive agriculture led to the stabilisation of many breeds as distinct genetic entities through the establishment of breed societies that defined breed characteristics and purity. Breeds very often originated from a regional population that contained diverse genetic origins. Breeds better adapted to modern production systems became more widespread, while other breeds consequently declined and even became extinct in a considerable number of cases. In the middle to late 20th century, modern within-breed genetic improvement programs became widely established. This was coupled with specialization in the livestock sector, extensive use of cross-breeding and the rise of breeding cooperatives and companies. Introduction of exotic genetic material continues to be seen as a solution to low productivity of local breeds even in areas where the exotic genotypes are ill-adapted. Economic globalization and government or donor subsidies make it relatively easy for farmers to acquire exotic germplasm. Change of species or breeds may offer more flexibility and quicker adaptation to change than making use of within breed variation only. In the short term, cross-breeding with improved breeds can show quicker impact than the time consuming and expensive selective breeding of a local breed. However, improved management and the recognition of non-market benefits in the productivity assessment show that local/indigenous breeds can outperform crossbreds under certain circumstances (for example, we can see Ayalew et al., 2003). Additionally, there were also a number of well documented examples of unsustainable livestock development projects based on importations of exotic genetic material in developing countries (e.g. Sainath, 1996).

2.5.1 Methods of breeding strategies undertaken by the Livestock keepers / breeders

Different strategies for breed development and breed improvement can be identified. Selection of preferred animals (male and female) for further multiplication is the basic method applied by livestock keepers and breeders. In modern breeding programmes this is complemented by estimation of breeding values and dissemination of superior genetic material. The more detailed the knowledge about the genetic background of desired traits, the more targeted this process can be. For livestock keepers in developing countries 'community-based management' of AnGR is relevant, in which case within breed selection is practiced where phenotypically selected male animals are rotated or exchanged between herds. Reproductive technologies have revolutionized the animal breeding sector and facilitated exchange of genetic material between countries and regions of the world. Breeders realize that these technologies have led to reduced genetic variation, mainly through the widespread use of a small number of superior animals. Molecular biology is also having an increasing impact on the animal breeding sector, as well as playing a role in the introduction of the patenting of processes and products used in animal breeding. Two recent studies (Valle Zárate et al., 2006; Mathias & Mundy, 2005) analyzed historical and current AnGR exchange and aimed to quantify and assess trends in the transfer of AnGR. There are several reasons why there is little baseline quantitative data on AnGR exchange. First of all, import and export figures of live animals, originating from veterinary or economic records, do not usually distinguish breeding animals from slaughter animals/animal products, and the origin of import and destination of export is often unclear. Secondly, multinational companies provide no data on the intra-company exchanges of genetic material between countries. Thirdly, in many parts of the world unrecorded livestock movements take place, e.g. due to transhumance.

2.5.2 The climate and the Altitude

In general, climate, and rainfall in particular, affects animals more than most other environmental factors. Breeds respond quite differently to severe conditions of humidity and temperature. Africa can be subdivided into five major agro-ecological zones based on annual rainfall (with the exception of the highland areas); arid (< 600 mm), semiarid (> 600 and < 1 000), sub-humid (> 1 000 and < 1 500), humid (> 1 500) and highland. Each supports a different type of vegetation and makes different demands on livestock. Many long-established breeds, including the zebu cattle, West African Dwarf sheep and goats and dromedaries, have evolved the necessary physiology to cope with hot climates. For example, sheep in the tropics are usually hairy whereas those in colder regions such as the highlands of Ethiopia are woolled. Major changes in climate or vegetation can cause quite dramatic changes in the composition of livestock. The encroachment of the Sahara into West Africa has resulted in the recent decline of a large number of indigenous breeds (the indigenous shorthorn taurine cattle breeds) and has brought others close to extinction. Within the major climatic zones some highly specific micro-environments exist which make unique sets of demands on animals, an example being the Lake Chad basin. Here the Kuri cattle have evolved morphologically (they are quite tall) and physiologically (they are resistant to endemic diseases) to cope with an island existence.

2.5.3 Infectious Diseases

In addition to the challenges created by the tropical climate and breeding strategies, many of the warmer regions harbour a range of infectious diseases that present a serious threat to livestock. Over one-third of Africa is infested by tsetse fly and consequently trypanosomiasis, for which it serves as a vector, is endemic. Animals indigenous to these regions are generally trypan tolerant, which has ensured their survival. For example, the N'Dama cattle of West Africa, although small in size and stature, possess an innate tolerance to the disease. In the past there have been large epidemics of other diseases, such as Rinderpest, which toward the end of the nineteenth century, wiped out nearly 90 percent of all African cattle. Rift Valley Fever, which affects small ruminants such as sheep and goats,and to a lesser extent cattle, and tick-borne diseases such as theileriosis, anaplasmosis, babesiosis and cowdriosis are all endemic in Sub-Saharan Africa. Breeds respond quite differently to such challenges. For example, the Ankole cattle breed is thought to be quite

2.6 Loss of farm animal genetic diversity and Genetic drift

There are more than 40 species of animals that have been domesticated (or semi-domesticated) during the past 10 to 12 thousand years that contribute directly or indirectly to agricultural production (FAO, 2000). Livestock have been undergoing constant genetic change since first domesticated. In contrast to many plant species, there is only a very limited genetic resource-base in wild populations of animal species that is of interest for farm animal breeding. The status of AnGR or livestock breeds, particularly in the developing world, is poorly understood. Loss of genetic diversity through the disappearance of local livestock breeds is widely reported but difficult to quantify. Hall & Ruane (1993) estimated that some 16% of the uniquely adapted breeds are extinct. FAO estimated (FAO, 2000) that about one third of all domesticated animal breeds were threatened with extinction. These figures have been criticised because they are based on national endangerment estimates, regardless of the total population size of the breed in question across countries and the genetic relationship between breeds across countries. After recent updates of the global Domestic Animal Diversity Information System (DAD-IS), the number of recognized domesticated animal breeds totals 7,600, consisting of 'transboundary breeds' (regional and international) and 'local/national breeds' (Scherf et al., 2006).

The term 'breed' does not have a universally accepted biological or legal definition. In developed countries breeds are characterised by clear definitions, physical characteristics and strict definitions of purity of pedigree, typically regulated by a breed society which is backed by law. In developing countries a breed is commonly defined by local tradition, identifying physical characteristics, a geographical location or ethnic group by which it was developed. On the other hand, Genetic drift that represents the random change of allele frequencies over generations due to finite population size, and a small Ne expected to increase the rate of genetic drift and associated loss of genetic variation across generations as alleles become fixed (Wright, 1929; Falconer and Mackay, 1996; Willi et al., 2006; Merilä, 2014). The high heterozygosity values sometimes reported in domestic breeds despite low Ne may, at least in part, result from biased selection of hypervariable microsatellite markers in chromosome regions not under selection (Taberlet et al., 2008). Kim et al. (2013) provide support for this hypothesis showing that Holstein dairy cattle selected intensively for increased milk yield do have increased overall autozygosity when assessed across the genome; this is likely a consequence of both genetic drift and selection.

Breeding in commercial dairy cattle breeds like Holstein Frisian and Jersey with Ne's below 100 has been very efficient and no apparent signs of selection plateaus have been seen (Chikhi et al., 2004; Sorensen et al., 2005; Hill and Kirkpatrick, 2010; Leroy et al., 2013). With very high selection intensity, most traits can be changed through directional selection (Hill and Kirkpatrick, 2010) and genetic variation may remain relatively stable within the time frames typically considered by a breeding company or an individual farmer. Many domestic species have long generation length, and intense selection within domestic breeds has (from an evolutionary point of view) not been practiced for long. The breed concept is not more than approximately 200 years old, and reproductive technologies enabling intense selection has only been in common use since the 1960s (Taberlet et al., 2011)

2.7 Current State of AnGR and Extinction

It is reported that the status of AnGR is poorly understood and loss of genetic diversity is difficult to quantify [FAO, 2006]. But, still there are salient facts that animal genetic resources are disappearing rapidly worldwide. For instance, from the existing 7,616 animal breeds, the status of 36% of breeds is neither known nor comprehensive genetic characterization is yet done [FAO, 2009]. Few reports also showed that from the existing breeds about 1000 breeds have extinct (Table I) during the last 100 years [I. K. Rollefson, 2005]. Currently, about one-fifth of the world's domestic livestock are at risk [FAO, 2012] and 10% are already extinct [G. Cicia, E. D. Ercole, and D. Marino, 2003]. Most of these breeds are from developing countries [M. Zerabruk, O. Vangen, and M. Haile, 2007] and it is also anticipated that the hotspots of breed loss and genetic erosion in the coming years will be in developing countries B. R. Marti, H. Simianer, J. Gibson, O. Hanotte, and J. E. O. Rege, 2003]. For instance, these studies reported higher extinction probability for some selected cattle breeds of cattle (Table II).

Species	Africa	Asia	Europe and caucasus	Latin America and carbean	Near and Middle east	North America	Southwest pacific	International trans boundary breeds	World
Ass	1	0	3	0	1	0	0	0	5
Buffalo	0	0	1	0	0	0	0	0	1
Cattle	20	18	120	19	1	1	2	1	182
Goat	0	2	15	0	0	1	0	0	18
Horse	6	1	72	0	0	8	1	0	88
pig	0	15	91	2	0	0	1	0	109
Rabbit	0	0	0	0	2	0	0	0	2
Sheep	5	6	144	0	1	1	2	0	159
Chicken	0	5	51	0	0	1	0	0	57
Total	32	47	497	21	5	12	6	1	621

Table I: Number of Extinct Livestock Breeds [7]

No.	Breed	Extinction probability	References
1	Sheko	0.77	Resit-Marti et al., 2003
2	Highland zebu	0.77	Resit-Marti et al., 2003
3	Begait	0.67	Zerabruk et al., 2007
4	Abergelle	0.53	Zerabruk et al., 2007
5	Bale	0.57	Resit-Marti et al., 2003
6	Irob	0.57	Zerabruk et al., 2007
7	Arsi	0.53	Resit-Marti et al., 2003
8	Arado	0.50	Resit-Marti et al., 2003
9	Ethiopia Boran	0.48	Resit-Marti et al., 2003
10	Abigar	0.47	Resit-Marti et al., 2003
11	Fogera	0.43, 0.47	Resit-Marti et al., 2003; Zerabruk et al., 2007
12	Afar	0.43, 0.47	Resit-Marti et al., 2003; Zerabruk et al., 2007
13	Raya	0.47	Zerabruk et al., 2007
14	Arado	0.37	Zerabruk et al., 2007

Table II: Extinction Probability of Some Selected Cattle Breeds of Ethiopia

3. Conservation strategies

Conservation strategies can be categorized as conserving animal's in-situ, in the environment in which they were developed, or ex-situ, all other cases. The latter can be further divided into ex-situ in vivo conservation and cryogenic storage. The CBD in its Article 8 gives clear priority to in-situ conservation and refers to it as the recovery and maintenance of species or breeds in the environment in which they have developed. This strategy is also the most preferred as the animals continue to evolve in their original habitat. Article 9 considers ex-situ conservation as: (i) the maintenance of small, closely managed populations outside their adapted environment in artificial or semi artificial settings; and (ii) freezing (cryoconservation) of genetic material such as semen, embryos, DNA, cells or ova. Ex-situ cryoconservation does not allow for further evolutionary progress of breeds which they would have undergone in their natural environment. In-situ and ex-situ conservation is complementary, not mutually exclusive. The decision will depend on a thorough evaluation of the situation and the possibilities to use one or the other strategy. For example, it is important to note that frozen germplasm can play an important role in support of in vivo animal conservation strategies.

3.1 Objectives for AnGR conservation

The objectives for AnGR conservation include economic, social and cultural, environmental, risk reduction, research and training. These objectives have been summarized elsewhere (Hodges,

1987; Henson, 1992). Animal diversity should be maintained for its economic potential in allowing responding quickly and swiftly to changes in market conditions, consumer preferences or environmental conditions; Animal diversity has an important social and cultural role. Animals are integral part of ceremonies and habits of social ethnic groups. In modern societies they provide recreation possibilities. Farm parks can serve as teaching aid for the urban. The tourism industry may be important in many countries, and may rely on the specific environment of which the local domestic breeds are an integral part; Animal diversity is an integral part of an agro eco-system. The loss of this diversity would contribute to higher risk in the production system, reduced ability to respond to change, degradation of the environment in question and could ultimately lead to its destruction. Marginal areas and low to medium input production systems, and increased integration of livestock into agricultural production will be important for food production in the developing world.

Maintenance and development of adapted breeds are of critical importance to ensure this can be achieved sustainably without adverse environmental impact; Indigenous animal diversity is important insurance to enable response to possible, but yet unknown, requirements in the future. It is risky to rely on only a few breeds: a concentration on a small number of breed's results in losses of genes and gene combinations which are not relevant at present, but which could become relevant in the future. Conserving domestic animal diversity is reducing the risk and enhancing food security. At issue is not only the possible accentuated loss of diversity but the lack of readily available gene combinations, particularly for adaptive fitness to specific environments when AnGR are lost; Animal diversity should be conserved for research and training purpose including basic biological research in immunology, nutrition, reproduction, genetics and adaptation to climatic and other environmental changes. Genetically distant breeds are needed for research into disease resistance and susceptibility helping to a better understanding of the underlying mechanisms and to the development of better treatments or management of the disease. The conservation activity serves as training for all stakeholders and this in turn leads to greater awareness, knowledge and reduced risk.

3.1. The breeds to conserve, priorities, and Categorization of Conservation

The Weitzman (1992; 1993) has presented a formal framework for the decision making process on breeds to conserve. However, the diversity metric suggested by Weitzman has been criticized as not accounting for within-breed diversity (see, for example, Caballero and Toro, 2002), the general framework to make decisions based on the expected conserved diversity has a strong appeal.

The Weitzman (1993) suggests that the conservation potential is the single most informative criterion to rank breeds with respect to conservation priority. The conservation potential of a breed basically reflects the amount of expected diversity that can be conserved if a breed is made completely safe. Simianer *et al.* (2003) found that this criterion correctly identified the optimum subset of six breeds to be conserved in a set of 23 African cattle breeds. Pinent, Simianer, and Weigend (2005) show that the derivation of the conservation potential for a set of breeds should always take into account information on related breeds outside the set of candidate breeds for conservation (e.g. foreign breeds or commercial breeding populations) to avoid "false positives". Therefore, Simianer (2002; 2005) suggests combining expected diversity with other criteria resulting in the expected total utility as a maximization criterion. This criteria may encompass the presence of special genetic traits (such as disease tolerance), production, and cultural or environment values of breeds, *inter alia*. A similar but less formal argument was made by Piyasatian and Kinghorn (2003).

Consideration of which refined conservation projects are needed and what the options are feasible to accomplish in vivo conservation for a breed. To do this there is a need to prioritise and select the appropriate conservation strategy. The primary categorisation is between in-situ and ex-situ conservation and are the two programs employed in conservation pro-grams of animal genetic resources. Ex situ conservation means conservation of animal genetic re-sources away from its original production systems where they were developed or are now normally found and bred [FAO, 2013] and Kasso and Balkrishnan, 2013]. This is maintenance of live animals in a zoo (ex situ in vivo) and cryopreservation of genetic material like semen, oocytes, embryos and DNA [Ibid]. In situ conservation involves production of animals in their original production environment either on-farm or community based and includes both actual farms and pastoral production systems.

In practice, ex situ in vivo conservation program suffers from disadvantages of variation of herd management on the farms from management of the herd in the field. Unlike herds under farmers' management, animals in the station may be spared migration, drought, and diseases and subjected to a different pattern of evolutionary processes (Adebabay et al, 20016). This means natural selection is usually no longer effective in its role of ensuring the adaptation of the population [Kasso and Balkrishnan, 2013]. At times numbers of animals in the zoo/ranch may be too few to represent the full diversity of the breed and the animals may become secluded from the wider gene pool. Furthermore, they are subjected to gradually change their characteristics in adaptation to their new environment. Similarly, the limitation of conservation of animal genetic resources in a gene bank is that it does not possess the breed's socioeconomic role, nor does it save its cultural, historical and ecological values. Besides, ex situ conservation requires appropriate infrastructure and organisation, technical capacity, legal arrangements and sustained funding [Fernandez, V. Beatriz, P. Ricardo and A. Miguel, 2005]. These days, the issue of budget is also becoming critical to run conservation programs in developing countries. The other question is this kind of conservation programs requires sufficient grazing land to sustain the existing animals. Nonetheless, to date land scarcity is becoming a pronounced challenge in many developing countries.

For several reasons, developing countries do not put conservation of animal genetic resources as a priority, mainly be-cause their main goals are increased production and competitiveness in the global market in the short term. Unfortunately, there are very few prospective efforts directed towards thinking about the future of genetic resources and breeding programs. Kenya has better experience than the other countries with improvement schemes (livestock recording and genetic evaluations) for all exotic dairy breeds and for some local beef or dual-purpose breeds like the Boran and Sahiwal cattle breeds as of Adebabay et al, 2016. In Ethiopia, from the 1950s to 1970s, conservation pro-grams were established in the form of ranches and multiplications centres for the conservation of Fogera, Boran, Horro and Arsi cattle breeds and Menz sheep [FAO, 2007] and [E. Workneh, Getahun, M. Tibbo, Y. Mamo, and J. E. O. Rege, 2004]. These include Metekel cattle ranch, Andassa cattle ranch, Wolaita cattle ranch, Jigjiga ogaden cattle ranch, Dida Tuyura Boran cattle and Abernossa Boran cattle ranch, Bako sheep ranch and Menz sheep ranch. However, most of them, including sheep ranches (Horro sheep ranch at Bako, Menz sheep ranch at Sheno and Amed-Guya menze sheep multiplication centre) are closed with a disappearance of thousands of animals. Besides, semen of both exotic and indigenous cattle breeds is stored in semen banks, but not regularly used. In Tanzania breeding programmes exist for Mpwapwa and Boran cattle breeds at research stations and for goats, breeding strategies exist for pure breeding of Blended, Newala,

Ujiji and Gogo breeds. In Uganda, breeding schemes are practised within research and development programmes for Ankole cattle as well as for some other cattle, goat and sheep breeds. In Zambia, characterization and conservation programmes are undertaken for some indigenous cattle, like Angoni, Barotse, Tonga and Baila breeds. Indigenous cattle breeds are being conserved in vivo at government stations

Table III. Number of Countries with Conservation Programs [13]

gion	No. of countries with in vivo conservation	No. of countries with in vitro conservation
rica	18	9
ia	13	12
rope and Caucasus	33	12
tin America and the Caribbean	8	6
ar and middle east	1	0
rth America	2	2
uth west pacific	2	1
orld	77	42

3.2 Priorities among strategies of active conservation

The Convention on Biological Diversity lays clear guidelines for the priorities among the strategies. Article 8 advocates the development of in-situ conservation activities where appropriate, whilst Article 9 calls for the development of exsitu conservation schemes but primarily as a complement to Considering the Options in-situ efforts. Therefore the Convention identifies that *in situ* conservation as the method of choice wherever possible. Integral to the definition of in-situ is the concept that the breed should continue to occupy the environment in which it was developed. Such a change would conserve the breed within its own locality, and so it would continue to have a relationship with its indigenous keepers and this must be considered of greater value than complete removal of the breed from the locality.

Conservation schemes can be envisaged that maintain live animals both in-situ and ex-situ and involve the regular interbreeding of the subgroups. Strictly speaking such schemes should be considered as ex-situ live animal conservation, but common sense must tell us that these have some higher quality than other ex-situ options. Cryoconservation must be considered as the strategy with the lowest priority.

4. Global Strategy for the Management of Farm Animal Genetic Resources

In recognition of the importance of animal genetic resources (AnGR), and of the sizeable portion which is currently at risk of loss, and in keeping with FAO's mandate and the Convention on Biological Diversity (CBD) a special action programme for the Global Management of Farm Animal Genetic Resources was launched by FAO in 1992. This Programme has as its objective to establish practical mechanisms and set of key actions by countries that aim, in particular, at: developing and making better use of animal genetic resources adapted to the world's major medium-input and low-input production environments, so as to enable their agricultural systems to intensify sustainably; and overcoming the serious threat of genetic erosion amongst the remaining 5,000 or so remaining breed resources of the 14 main farm animal species.

The framework of the Programme comprises four basic components: An inter-governmental mechanism whereby governments can directly guide international policy development, within the Commission on Genetic Resources for Food and Agriculture; A global, country-based structure with three elements: (i) focal points and networks, including the provision of a National Focal Point responsible for implementing and maintaining the networks within the country, and for undertaking technical exchanges with FAO on the Global AnGR programme; (ii) a stakeholders mechanism to properly involve the broad Introduction of parties; and (iii) use of the country-secure Domestic Animal Diversity Information System (DAD-IS); A programme of technical activities, of six elements: (i) characterization; (ii) in-situ utilization and conservation; (iii) in-situ and ex-situ conservation; (iv) guidelines and action planning; (v) the development of a communications and information system, and relevant training; and (vi) coordination; Expert cadres to guide development of the strategy, and maximize the cost-effectiveness of country participation.

As mentioned above, one of the objectives of this Programme is the development of Guidelines for country use. The Primary Guideline Document (FAO, 1996), mainly targeted towards policy makers, is designed to help countries get started to identify the main elements and objectives of an animal genetic resources management plan, and to outline the strategic policy directions required to fulfil these objectives.

5. Conclusions and future scenarios

AnGR comprises all animal species, breeds and strains that have economic, scientific and cultural value to mankind in terms of food and agricultural production for the present and the posterity [J. Gibson et al, 2005]-[J. E. O. Rege and A. M. Okeyo, 2006]. For the last dozen thousand years, about 40 animal species have been tamed or semi-tamed worldwide [FAO, 2013]. In developing countries, animal genetic resources (AnGR) are a very crucial component of biodiversity [FAO, 2003] nourishing 70% of the world's rural poor. These comprised of 194 million pastoralists, 686 million mixed farmers, and 107 million landless livestock keepers [FAO, 2007 13]. The effort to improve food security in developing countries lies in wise use of genetic diversity [J. Philipsson, J. E. O. Rege, E. Zonabend, and A. M. Okeyo , 2011 14]. The values of AnGR conservation are mentioned in enormous literatures. All of them entirely appraise the past and present contribution of animal genetic resources to people under different environmental conditions [J. Philipsson, J. E. O. Rege, E. Zonabend, and A. M. Okeyo, 2011], [B. M. A. O. Perera, 2010]. However, animal genetic resources are depleting for various defined reasons in developing countries [FAO, 2008] and [J. E. O. Rege and J. P. Gibson, 2008]. The great concerns are the inflated loss of indigenous breeds impacting the livelihood options for the poor owing to utilization and management of these genetic resources [C. Tisdell, 2003]. There are several factors which place breeds at risk of loss and threaten domestic animal diversity. By far the greatest cause for genetic erosion is the growing trend to global reliance on a very limited number of modern breeds suited for the high input-output needs of industrial agriculture. These include, i) cross-breeding with and/or replacement by imported breeds in programmes designed to improve animal productivity, ii) neglect arising from shifts in social settings, production systems and/or market demand of certain animal products, iii) urbanization and its impact on traditional animal agriculture, iv) drought, v) civil strife/conflicts and vi) famines. The loss of breeds is only one indicator for the loss of genetic diversity in farm animal species, as Hammond & Leitch (1996) observe that the genetic variance between breeds accounts for approximately 30-50% of the total variance. Very little is known about the distribution of specific genes and the genetic uniqueness of breeds, although a range of rather unique adaptation traits have been documented for local/indigenous breeds. It has estimated that if the due attention is not given to biodiversity a massive of AnGR will be loss in upcoming 20 years and proper record of breeds for all Animal Species is important to follow them up (ranching, allocate proper land and resources ..). As a future scenarios in one way or another and may affect exchange, use and conservation of AnGR. The on-going globalization process will certainly affect exchange patterns and will negatively affect conservation of farm animal genetic diversity. Diseases and disasters have happened but are unpredictable. It is clear however, if such a scenario happens, this could seriously threaten AnGR. The effects of biotechnology and climate change were generally considered as long term.

References

FAO. 2000. *World watch list for domestic animal diversity.* 3rd edition. Rome. (also available at dad.fao.org/en/refer/library/wwl/wwl3.pdf)

L. F. Groeneveld, J. A. Lenstra, H. Eding, M. A. Toro, B. Scherf, D. Pilling, et al., "Genetic diversity in farm animals: A review," International Society for Animal Genetics, Animal Genetics, vol. 41, pp. 6-31, 2010.

FAO (2009a) Status and Trends Report on Animal Genetic Resources – 2008. CGRFA/WG-AnGR-5/09/Inf. 7. Available at: http://www.fao.org/ag/AGAInfo/programmes/en/genetics/documents/ITWG_AnGR_5_09_3_2.pdf

Convention on Biological Diversity (CBD) (2009): Connecting biodiversity and

climate change mitigation and adaptation. Report of the Second Ad Hoc Technical Expert Group on Biodiversity and Climate Change, Montreal, Technical Series

No. 41, 126 pp. Available at: https://www.cbd.int/doc/publications/cbd-ts-41-en.pdf.

Hoffmann, "Climate change and the characterization, breeding and conservation of animal genetic resources," *International Society for Animal Genetics, Animal Genetics, vol. 41, pp. 32-46, 2010*

S. Anderson, "Animal genetic resources and sustainable livelihoods," Ecological economics, vol. 45, pp. 331-339, 2003.

Ayalew W., King J.M., Bruns E. & Rischkowsky B. (2003) Economic evaluation of smallholder subsistence livestock production: lessons from Ethiopian goat development program. *Ecological Economics* **5**, 473– 85.

FAO, 1999. The Global Strategy for the Management of Farm Animal Genetic Resources. FAO, Rome, Italy

Anderson, S., 2003. Animal Genetic Resources and Sustainable Livelihoods. Ecological Economics, forthcoming.

Steinfeld H., Wassenaar T. & Jutzi S. (2006) Livestock production systems in developing countries: status, drivers, trends. *Revue Scientific et Technique office International des Epizooties* **25**, 505– 16.

A. S. Mariante, M. S. M. Albuquerquea, A. A. Egito, C. McManus, M. A. Lopes, and S. R. Paiva, "Present status of the conservation of livestock genetic resources in Brazil," Livestock Science, vol. 120, pp. 204-212, 2009.

Gollin D., Van Dusen E. & Blackburn H. (2008) Animal genetic resource trade flows: economic assessment. *Livestock Science* **120**, 248– 55.

Hammond, K. and H. Leitch, 1996. The FAO Global Programme for the Management of Farm Animal Genetic Resources. In: Miller R., V. Pursel, and H. Norman (eds.), Beltsville Symposia in Agricultural Research. XX. Biotechnology's Role in the Genetic Improvement of Farm Animals. American Society of Animal Science, IL: 24-42.

FAO, "Animal genetic resources conservation and development: The role of FAO," Arch. Zootec., vol. 52, pp. 185-192, 2003

Zohrabian, A., G. Traxler, S. Caudil, and M. Smale, 2003. Valuing Pre-Commercial Genetic Resources: A Maximum Entropy Approach. American Journal of Agricultural Economics, 85: 429-436

Mendelsohn, R., 2003. The Challenge of Conserving Indigenous Domesticated Animals. Ecological Economics, forthcoming.

FAO (2009a) Status and Trends Report on Animal Genetic Resources – 2008. CGRFA/WG-AnGR-5/09/Inf. 7. Available at: http://www.fao.org/ag/AGAInfo/programmes/en/genetics/documents/ITWG_AnGR_5_09_3_2.pdf.

Eding H. (2008) A breed is a breed if enough people say it is. *Editorial, Globaldiv Newsletter* **4**, 1– 4.

FAO (2009c) Threats to Animal Genetic Resources – Their Relevance, Importance and Opportunities to Decrease Their Impact. CGRFA Background Study Paper 50, 55 pp. Available at: ftp://ftp.fao.org/docrep/fao/meeting/017/ak572e.pdf.

J. Gibson, S. Gamage, O. Hanotte, L. Iñiguez, J. C. Maillard, B. Rischkowsky, D. Semambo, and J Toll, "Options and strategies for the conservation of farm animal genetic resources," Report of an International Workshop, Montpellier, France, Nov. 2005

Valle Zárate A., Musavaya K. & Schäfer C. (2006) *Gene Flow in Animal Genetic Resources. A Study on Status, Impact and Trends*. Institute of Animal Production in the Tropics and Subtropics, University of Hohenheim, Germany.

Hall S.J.G. (2004) *Livestock Biodiversity. Genetic Resources for the Farming of the Future.* Blackwell, Oxford, 264 pp.

Wright, S. (1931). Evolution in mendelian populations. *Genetics* 16, 97–159.

Willi, Y., Van Buskirk, J., and Hoffmann, A. A. (2006). Limits to the adaptive potential of small populations. *Annu. Rev. Ecol. Syst.* 17, 433–458. doi: 10.1146/annurev.ecolsys.37.091305.110145

Taberlet, P., Valentini, A., Rezaei, H. R., Naderi, S., Pompanon, F., Negrini, R.,et al. (2008). Are cattle, sheep, and goats endangered species? *Mol. Ecol.* 17, 275–284. doi: 10.1111/j.1365-294X.2007.03475.x

Kim, E.-S., Cole, J. B., Huson, H., Wiggans, G. R., Van Tassell, C. P., Crooker, B. A.,et al. (2013). Effect of artificial selection on runs of homozygosity in US Holstein cattle. *PLoS ONE* 8:e80813. doi: 10.1371/journal.pone.0080813

Chikhi, L., Goossens, B., Treanor, A., and Bruford, M. W. (2004). Population genetic structure of and inbreeding in an insular cattle breed, the Jersey, and its implications for genetic resource management. *Heredity* 92, 396–401. doi: 10.1038/sj.hdy.6800433

Sørensen, A. C., Sørensen, M. K., and Berg, P. (2005). Inbreeding in Danish dairy cattle breeds. *J. Dairy Sci.* 88, 1865–1872. doi: 10.3168/jds.S0022-0302(05)72861-7

Hill, W. G., and Kirkpatrick, M. (2010). What animal breeding has taught us about evolution. *Annu. Rev. Ecol. Evol. Syst.* 41, 1–19. doi: 10.1146/annurev-ecolsys-102209-144728

Taberlet, P., Coissac, E., Pansu, J., and Pompanon, F. (2011). Conservation genetics of cattle, sheep, and goats. *C. R. Biol.* 334, 247–254. doi: 10.1016/j.crvi.2010.12.007

FAO, "Exchange, use and conservation of animal genetic resources: Policy and regulatory options," Centre for Genetic Resources, the Netherlands (CGN) Report, Rome, Italy, 2006.

FAO, "The use and exchange of animal genetic resources for food and agriculture," Background Study Paper no. 43, 2009

K. Rollefson, "Building an international legal framework on animal genetic resources: Can it help the dry lands and foodinsecure countries? League for Pastoral Peoples," German NGO Forum on Environment & Development, Bonn, Germany, 2005.

FAO, "Cryo-conservation of animal genetic resources," FAO Animal Production and Health Guidelines, No. 12. Rome, Italy, 2012.

G. Cicia, E. D. Ercole, and D. Marino, "Costs and benefits of preserving farm animal genetic resources from extinction: CVM and Bio-economic model for valuing a conservation program for the Italian Pentro horse," Ecological Economics, vol. 45, pp. 445- 459, 2003.

M. Zerabruk, O. Vangen, and M. Haile, "The status of cattle genetic resources in North Ethiopia: On-farm characterization of six major cattle breeds," Animal Genetic Resources Information, vol. 40, pp. 15-32, 2007.

B. R. Marti, H. Simianer, J. Gibson, O. Hanotte, and J. E. O. Rege, "Weitzman's approach and conservation of breed diversity: An application to african cattle breeds, conservation biology," vol. 5, no. 17, pp. 1299-1311, 2003.

Weitzman, M.L., 1992. On Diversity. Quarterly Journal of Economics, 107: 157-183. Weitzman M., 1993. What to Preserve An Application of Diversity Theory to Crane Conservation. Quarterly Journal of Economics, 108: 157-183

Simianer, H., S. Marti, J. Gibson, O. Hanotte, and J.E.O. Rege, 2003. An Approach to the Optimal Allocation of Conservation Funds to Minimize Loss of Genetic Diversity Between Livestock Breeds. Ecological Economics: forthcoming

FAO, "In vivo conservation of animal genetic resources," FAO Animal Production and Health Guidelines, No. 14, Rome, Italy, 2013.

Kasso and Balkrishnan, Ex Situ Conservation of Biodiversity with Particular Emphasis to Ethiopia, Review Article, Hindawi Publishing Corporation, 2013

J. Fernandez, V. Beatriz, P. Ricardo and A. Miguel, "Efficiency of the use of pedigree and molecular marker information in conservation programs," Genetics, vol. 170, pp. 1313-1321, 2005.

FAO, "The State of the World's Animal Genetic Resources for Food and Agriculture," B. Rischkowsky and D. Pilling, Eds. Rome, Italy, 2007

E. Workneh, Getahun, M. Tibbo, Y. Mamo, and J. E. O. Rege, "Current state of knowledge on characterisation of farm animal genetic resources in Ethiopia. In: farm animal biodiversity in

Ethiopia: Status and prospects," Proceedings of the 11th annual conference of the Ethiopian Society of Animal Production (ESAP), Addis Ababa, Ethiopia, August 28-30, 2004

J. Gibson, S. Gamage, O. Hanotte, L. Iñiguez, J. C. Maillard, B. Rischkowsky, D. Semambo, and J Toll, "Options and strategies for the conservation of farm animal genetic resources," Report of an International Workshop, Montpellier, France, Nov. 2005.

J. E. O. Rege and A. M. Okeyo, "Improving our knowledge of tropical indigenous animal genetic resources," in Animal Genetics Training Resource, J. M. Ojango, B. Malmfors, and A. M. Okeyo, Eds. International Livestock Research Institute, Nairobi, Kenya, and Swedish University of Agricultural Sciences, Uppsala, Sweden, 2006.

FAO, "In vivo conservation of animal genetic resources," FAO Animal Production and Health Guidelines, No. 14, Rome, Italy, 2013.

FAO, "The State of the World's Animal Genetic Resources for Food and Agriculture," B. Rischkowsky and D. Pilling, Eds. Rome, Italy, 2007.

J. Philipsson, J. E. O. Rege, E. Zonabend, and A. M. Okeyo, "Sustainable breeding programmes for tropical farming systems," in Animal Genetics Training Resource, J. M. Ojango, B. Malmfors, and A. M. Okeyo, Eds. International Livestock Research Institute, Nairobi, Kenya, and Swedish University of Agricultural Sciences, Uppsala, Sweden, 2011.

B. M. A. O. Perera, "The changing pattern of livestock farming systems in South-Asia, threats to indigenous breeds, and programs International Journal of Pharma Medicine and Biological Sciences Vol. 5, No. 1, January 2016 to conserve animal genetic resources," Animal Conservation Program in Emerging Economies, pp. 37-41, 2010.

FAO, Global Plan of Action for Animal Genetic Resources and the Interlaken Declaration, Commission on Genetic Resources for Food and Agriculture, Rome, Italy, 2008.

J. E. O. Rege and J. P. Gibson, Animal genetic resources and economic development: Issues in relation to economic valuation, Ecological Economics, vol. 45, pp. 319-330.

C. Tisdell, "Socioeconomic causes of loss of animal genetic diversity: Analysis and assessment," Ecological Economics, vol. 45, pp. 365-376, 2003.

YOUR KNOWLEDGE HAS VALUE

- We will publish your bachelor's and master's thesis, essays and papers

- Your own eBook and book - sold worldwide in all relevant shops

- Earn money with each sale

Upload your text at www.GRIN.com and publish for free